W9-CLH-267

TALKING POINTS

NATURE VS MAN

Antony Mason

Stargazer Books
Mankato • Minnesota

About This Book

NATURE vs MAN looks at the problem of natural disasters and their effects on mankind. Because our population is growing at such a fast rate, more people are killed every time a nation is hit by a disaster. Somehow we must learn to work with nature, not against it.

Produced by Aladdin Books Ltd

First published in 2009
in the United States by
Stargazer Books,
distributed by
Black Rabbit Books
P.O. Box 3263
Mankato, MN 56002

Printed in the United States

Designer: Flick, Book Design and Graphics
Editor: Katie Harker
Picture Researcher: Alexa Brown
The author: Antony Mason is a freelance editor and author of more than 60 books for children and adults.
The consultant: Rob Bowden is an education consultant, author, and photographer specializing in social and environmental issues.

Library of Congress Cataloging-in-Publication Data

Mason, Antony.
Nature vs. man / Antony Mason.
p. cm. -- (Talking points)
Includes index.
ISBN 978-1-59604-186-8
I. Title.
GB5019.M378 2009
304.2--dc22
2008016393

Contents

INTRODUCTION

Humans have always tried to control their environment, but nature's force is more powerful than anything that man has ever created. A natural disaster can flatten homes, kill people, and wreck lives. We need to learn how to cope with the forces of nature.

A MATTER OF STATISTICS

Humans tend to rate the severity of disasters in terms of the number of deaths and casualties. Statistics make headlines. If a volcano erupts in a remote area, it is not considered to be a worthy news item. However, if an earthquake takes the lives of thousands of people, the news will spread quickly. Because of the sheer number of people now living on our planet, we are likely to provoke certain kinds of disaster because of the daily impact we have on the environment.

Trying to explain natural disasters

For the individuals concerned, the scale of a disaster is not important, it is the loss of those closest to them that is the most devastating experience.

Because the world population is rising, the effects of nature are set to get worse in the future.

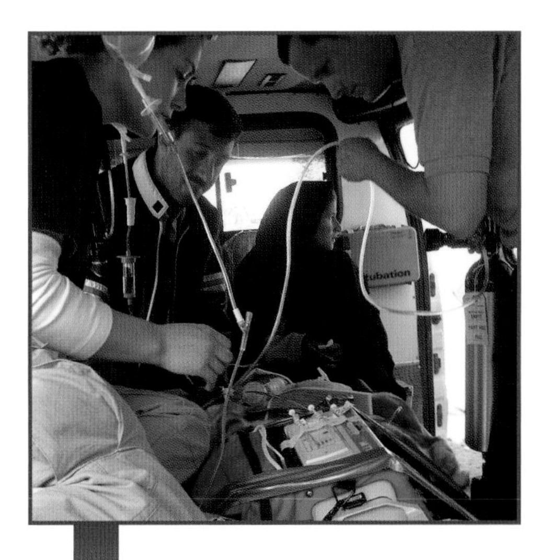

News of a natural disaster can spread very quickly. Sometimes, rescue teams attend to a scene within hours.

Natural disasters have always been a part of human life, but it is difficult for us to explain them and why they happen. In the past, people believed they were sent by their gods to teach them a lesson.

As scientific research has become more advanced, we are beginning to learn why certain natural disasters take place. This advanced technology also means that the news of the disaster can reach rescue teams much faster. Aid workers are often able to get to the scene of the disaster within hours.

Ten of the worst natural disasters since 1881

Approx. deaths	Type	Place	Date
1. 3 million	Flood	Huang He and Yangtze Rivers, China	1931
2. 1.5 million	Flood	Huang He and Yangtze Rivers, China	1887
3. 500,000	Cyclone/flood	Bhola, Bangladesh	1970
4. 300,000	Cyclone	Haiphong, Vietnam	1881
5. 280,000	Tsunami	Indonesia /Thailand / Sri Lanka / India	2004
6. 242,000*	Earthquake	Tangshan, Hebei Province, China	1976
7. 200,000	Earthquake	Qinghai, Gansu Province, China	1927
8. 200,000	Flood	Shanghai / Yangtze River, China	1911
9. 180,000	Earthquake	Xining, Gansu Province, China	1920
10. 160,000	Earthquake	Southern Italy and Sicily	1908

*This is the official figure, issued by the Chinese government. Outside sources suggest that the total may have been far higher, possibly 750,000 deaths.

These statistics do not include disasters in which deaths were caused by famine or epidemics alone.

What Is a Natural Disaster?

Any major event not caused by the actions of man is called a natural disaster. The event may injure or kill humans, or simply damage their property. If a volcano erupts on an uninhabited island, unless it affects humans or wildlife, it is not considered to be a natural disaster. The main natural disasters are caused by the weather or movement in the earth's crust.

Some of the greatest damage is caused by weather (*above*) and movement in the earth's crust (*left*).

The restless earth

Ever since its creation 4.5 billion years ago, the earth has been unstable. It only has a thin crust, which is constantly damaged by earthquakes and volcanic eruptions. Inside the earth are plates which push together and pull apart under massive pressure. This pressure may result in the ground shaking violently, as in an earthquake. The force of an earthquake is measured using the Richter scale.

If an earthquake occurs in a built-up area, it can cause massive devastation —houses will crumble, roads, bridges, and tunnels will collapse, and gas and water pipes will burst.

Tectonic plates

In the 1960s, geologists found a way to explain why earthquakes occurred more frequently in one place than the other. This is because the tectonic plates move about like giant rafts and in some places these plates are actually moving apart.

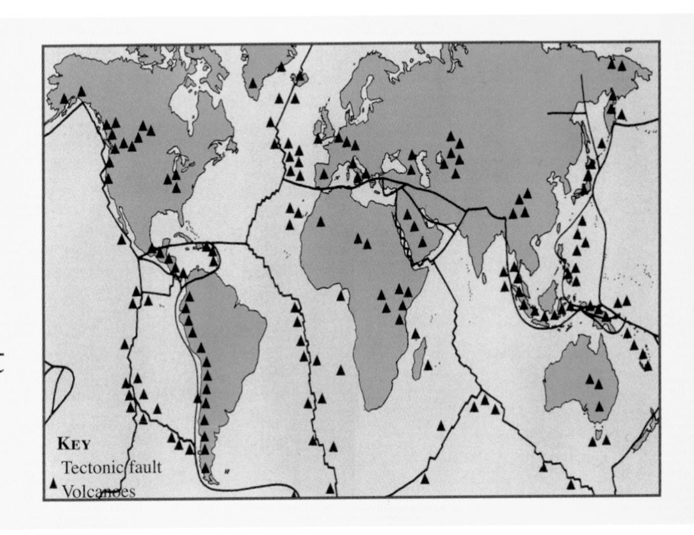

Volcanic eruptions

While the plates are moving, the liquid rock (or magma) beneath the earth's surface is also trying to force its way through the crust. When it finds a weak spot, a volcano occurs. Some volcanoes only release the pressure slowly, while others build up enormous pressure before erupting in a gigantic explosion. This kind of eruption happened at Vesuvius in AD79, burying the Roman cities of Pompeii and Herculaneum and killing some 3,500 people.

Volcanic eruptions can cause major devastation, like covering an entire region with volcanic ash and mud, as did the one in 1991 at Mount Pinatubo in the Philippines. In 1985, when the Nevado del Ruiz erupted in Colombia, a huge mudslide came hurtling down the mountain and engulfed a town. This eruption killed 21,000 out of the 23,000 inhabitants.

The eruption of Mount St. Helens in 1980 killed 57 people, destroyed 200 homes, 47 bridges, 15 miles (24 km) of railroad, and 186 miles (300 km) of road.

Tsunamis

When an earthquake or sudden volcanic eruption occurs near or beneath the sea, it can cause a massive tidal wave that sweeps across the water until it hits land. This is known as a tsunami. As a large number of the world's population live near the coast, the loss of life can be huge. In 1755, the capital of Portugal, Lisbon, was destroyed by an earthquake and the tsunami that followed half an hour later. People running from the earthquake thought it was safer to stand by the sea, unaware of the huge tidal wave heading their way.

Why are tsunamis so destructive?

Tsunamis cause major destruction because of the enormous volumes of water that are displaced. In the middle of the ocean the wave would be barely noticeable, but as it reaches the coast the sea becomes shallower and it creates friction against the seabed. This causes an enormous wave, which can rise as high as 330 ft (100 m). These waves can move as fast as 497 mph (800 kph).

The tsunami that hit Indonesia, Thailand, Sri Lanka, and India in December 2004 killed over 280,000 people. It was triggered by a powerful undersea earthquake.

THE RESTLESS SKIES

Weather is just as unpredictable, and can cause a range of natural disasters. Tropical cyclones (also known as hurricanes or typhoons), are vicious storms with winds of up to 199 mph (320 kph).

They are caused by a warm climate around the equator. When tropical cyclones hit the coast, they can throw boats out of harbors, tear through buildings, and even uproot trees. They are usually accompanied by heavy rainfall that can result in flooding. In fact about 90 percent of all deaths from tropical cyclones are caused by drowning. In 2005, Hurricane Katrina caused major flooding in the city of New Orleans, Louisiana—a city largely built below sea level.

Tornadoes

A tornado is another kind of deadly wind. It is caused by pressure building up in the atmosphere during thunderstorms. Tornadoes form funnels that spiral at speeds of up to 298 mph (480 kph). They can occur anywhere in the world, but the most famous region known for its tornadoes is "Tornado Alley" in the U.S. In 1974, a record number of 144 tornadoes hit the U.S. over a period of just two days. It killed 330 people and injured a further 5,000.

Cyclones can be 186 miles (300 km) wide and cause winds of 199 mph (320 kph).

Tornadoes are powerful funnels of wind that travel at speeds of up to 298 mph (480 kph).

FLOODING

Of all the weather-related natural disasters, flooding causes the most devastation and loss of life. It is very often the result of excessive rainfall, which can swell a river. On average, floods cause more deaths each year than any other natural disaster.

The world's worst floods occurred in Bangladesh and China. In 1970 alone, Bangladesh lost 500,000 lives from cyclone-induced flooding. The worst flooding in terms of human loss was in China in 1931, when the Huang He (Yellow River) broke its banks after heavy rain. Three million people died from drowning, famine, and disease.

Bangladesh: a special case

Hundreds of thousands of lives have been lost in regular disasters in Bangladesh over the last century. Strong winds and heavy rainfall frequently cause the country's many rivers to flood. Bangladesh is a poor country, but with help they are managing to build flood defenses to try and reduce the damage of future natural disasters.

Once a forest fire takes hold, it can burn for many weeks.

Fire

Long spells of hot, dry weather can often turn forests into huge areas of flammable material. It only takes one little spark and a little wind to stir up a major fire.

Forest fires can spread very quickly, traveling at 100 mph (160 kph). The flames can rise as high as 160 ft (50 m) at scorching temperatures of 3,632°F (2,000 °C). Very little, if anything, can survive these conditions. A forest fire is very difficult to put out and can burn for many weeks. The worst fire in recent history was in October 2003, in San Diego County. It lasted for ten days, destroyed 2,000 homes, killed 14 people, and caused damage estimated at $U.S.2 billion.

The Spanish Flu

The most devastating epidemic in history took place in 1918–19, when the world was hit by "Spanish flu" or influenza. When a disease is widespread it is known as "pandemic." This epidemic is believed to have killed between 20 and 40 million people.

The epidemic lasted for 18 months and the symptoms were so severe, many people died within 48 hours of showing the first signs. The exact cause of the epidemic is not known, but it was carried by soldiers and other people on the move after World War I. It affected virtually every country throughout the world.

ATTACKS ON HEALTH

Although disease in itself is not normally considered a natural disaster, in the aftermath of a disaster it can play a very devastating role. Epidemics such as cholera, caused from washing in and drinking contaminated water, can affect a widespread population. Such an impact can surely be classed as a natural disaster.

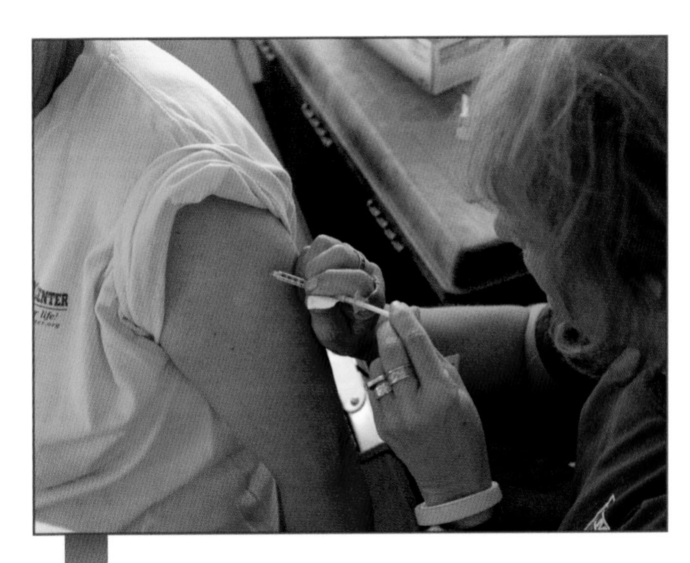

There are now many vaccinations against deadly diseases.

Can we fight the flu?

Influenza is a viral infection of the lungs that can cause fever, a cough, and severe muscle aches. It is serious for the elderly, very young, and people who suffer from conditions such as asthma.

Although it is possible to have a flu vaccination, the strain of the virus changes all the time. Scientists have to constantly work to create new vaccines. In 2003, an epidemic swept through Asia, revealing a new strain of the virus known as H5, which is normally associated with birds and not humans.

Epidemics of the past

The Black Death (1347–50) was one of the deadliest pandemics in human history. It killed around one third of Europe's population (25 million people) and similar numbers in Asia and Africa. It was named the "Black Death" because of the black spots that appeared on the body. Most people died within three days.

The Black Death and the Great Plague that hit England in 1665–66, are believed to have been caused by fleas carried by rats that were very common in towns and cities. The fleas bit into their victims, injecting them with the disease.

Today, because so many people travel abroad, they are being exposed to common diseases and therefore build up their own immunity. Isolated communities, however, are still vulnerable. Imported diseases such as measles, whooping cough, and flu, which were already common in Europe, hit the Americas in the 16th century and the native people had no natural defenses.

The spread of disease

There have, in the past, been hopes that many of these diseases could be totally wiped out. However, new and dangerous diseases continue to emerge, such as Acquired Immune Deficiency Syndrome (AIDS). It was first identified in 1981, and has now killed more than 25 million people. It is a sexually transmitted disease or, in the case of drug addicts, through the use of shared needles.

Another lethal disease is Ebola. It was first identified in 1976 as one of the most deadly diseases known to mankind. It causes death in 50–90 per cent of all cases. So far outbreaks of Ebola have been limited to Africa and the Philippines, but there are fears it could spread.

Diseases that change

Diseases are constantly changing and adapting to defeat modern medicine. One of the worst is malaria, which kills more than a million people every year. Malaria is found in the warmer parts of the world, and is caused by a parasite in female mosquitoes.

Other infections, like MRSA and C. difficile—superbugs found in hospitals—have also developed a resistance to antibiotics.

SARS

In 2002, a new disease similar to pneumonia emerged, called Severe Acute Respiratory Syndrome (SARS). This disease first appeared in China, but quickly spread. The disease caused a major scare worldwide, until it was brought under control a year later.

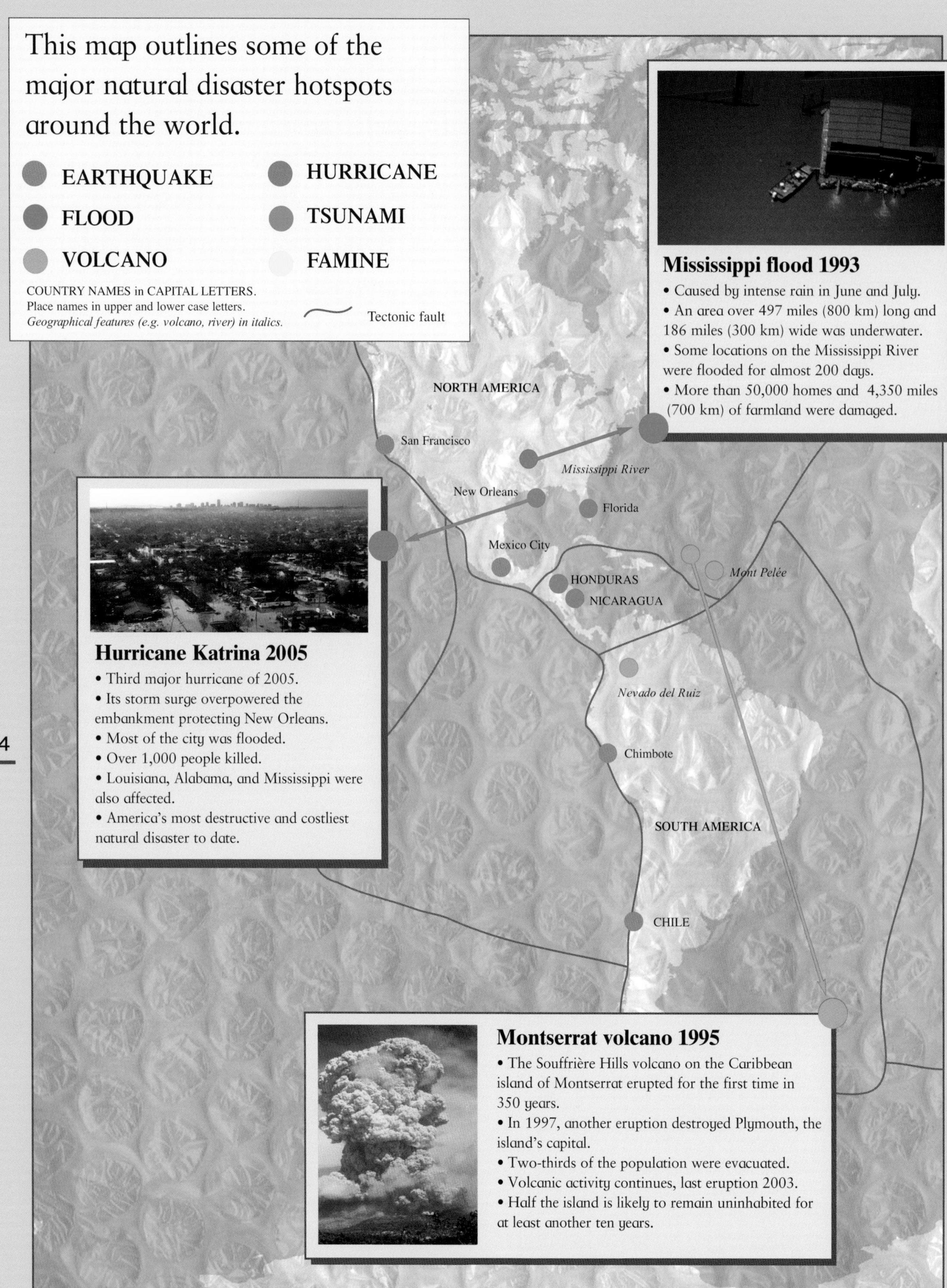

Mississippi flood 1993

- Caused by intense rain in June and July.
- An area over 497 miles (800 km) long and 186 miles (300 km) wide was underwater.
- Some locations on the Mississippi River were flooded for almost 200 days.
- More than 50,000 homes and 4,350 miles (700 km) of farmland were damaged.

Hurricane Katrina 2005

- Third major hurricane of 2005.
- Its storm surge overpowered the embankment protecting New Orleans.
- Most of the city was flooded.
- Over 1,000 people killed.
- Louisiana, Alabama, and Mississippi were also affected.
- America's most destructive and costliest natural disaster to date.

Montserrat volcano 1995

- The Souffrière Hills volcano on the Caribbean island of Montserrat erupted for the first time in 350 years.
- In 1997, another eruption destroyed Plymouth, the island's capital.
- Two-thirds of the population were evacuated.
- Volcanic activity continues, last eruption 2003.
- Half the island is likely to remain uninhabited for at least another ten years.

Bam earthquake 2003

- Earthquake measuring 6.5 on the Richter scale.
- 26,000 killed, 30,000 injured.
- 85 percent of the city's buildings were destroyed.
- Almost 75,000 left homeless.

Pakistan earthquake 2005

- Earthquake measuring 7.6 on Richter scale.
- Kashmir worst affected.
- Epicenter at Kashmir was close to the Earth's surface.
- Over 80,000 were killed.

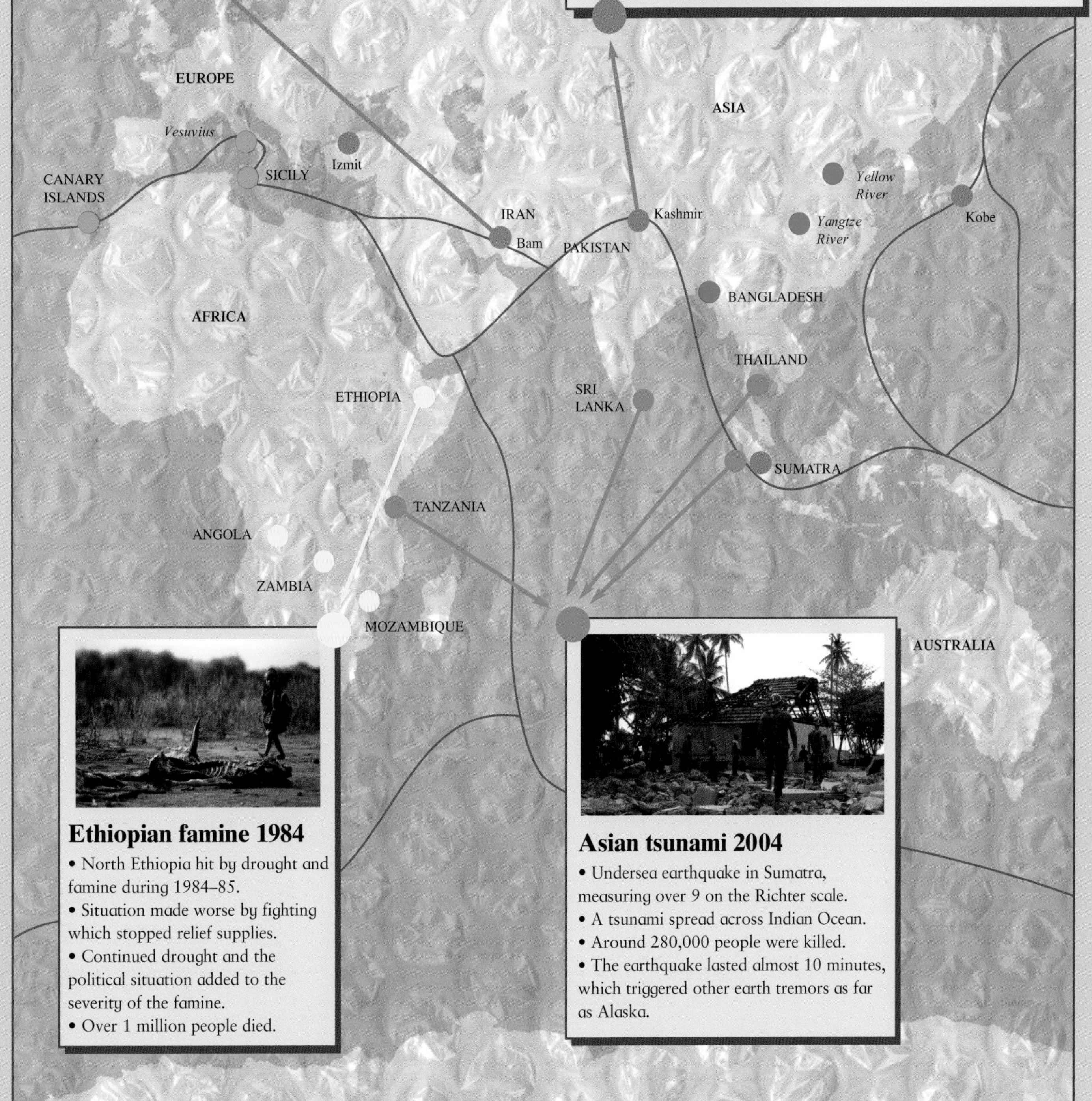

Ethiopian famine 1984

- North Ethiopia hit by drought and famine during 1984–85.
- Situation made worse by fighting which stopped relief supplies.
- Continued drought and the political situation added to the severity of the famine.
- Over 1 million people died.

Asian tsunami 2004

- Undersea earthquake in Sumatra, measuring over 9 on the Richter scale.
- A tsunami spread across Indian Ocean.
- Around 280,000 people were killed.
- The earthquake lasted almost 10 minutes, which triggered other earth tremors as far as Alaska.

Drought is a major cause of starvation in many parts of Africa and Asia.

ATTACKS ON FOOD

A natural disaster also occurs when people's food supplies are affected by a natural cause—as a result of drought, for example. If there has been no rain for a long time, the crops die and people run the risk of starvation.

People are reliant on nature to provide them with precious water. Those people who live in areas where drought is common try to find ways of storing food, or using alternative sources of water. However, if the country is hit by war, for example, this can add to their problems. When rain failed to fall in Ethiopia in the 1980s, around one million people died as a result. The worst recorded famine from drought was in China from 1876–9, when nine million people died.

Pests and diseases

It is not always rain that is the cause of famine. Sometimes crops can be hit by disease or pests. This happened in northern Africa in 2004, when a huge swarm of locusts ate all the crops in their path. It is estimated that it cost U.S.$122 million to try to stop the swarm of locusts from doing any further damage.

In Niger, in 2005, drought and locusts caused food shortages.

When Disaster Strikes

Some natural disasters can be predicted, giving people enough time to evacuate the area. Others, however, can strike suddenly. Those who are unlucky enough to be caught up in a natural disaster often need urgent medical attention. Added to their injuries is the stress of losing loved ones and all their personal belongings.

The injured always come first.

The immediate aftermath

Because some disasters—such as earthquakes—can take us by surprise, they often take many lives. Those who do survive have to face the loss of their loved ones, damage to their homes and work places, and disruption to local services. Sometimes the aftermath of a disaster can leave a lasting mental scar. Survival is the first priority of victims of disasters, who find their world collapsing around them. They often survive by quick thinking, or others simply by good fortune.

Rescue teams often face chaos when attending to a natural disaster.

Picking up the pieces

When the danger is past, the first task for survivors of a disaster is to try and find other survivors. People often get trapped when buildings collapse.

With modern communications, the world is quickly alerted to a natural disaster. This means that rescue teams from other countries can get to the scene as early as possible. International organizations use specialist teams to bring food, medical supplies, and experts such as doctors, nurses, firefighters, electricians, and builders to try and help the people rebuild their lives. Sometimes a political situation may hamper the aid workers, such as in Burma in 2008, when a government decides they do not want outside interference.

Misguided priorities

The Mont Pelée volcano gave plenty of warning before it erupted in 1902. There would have been time to evacuate the area, but the authorities decided to delay

because an election was in progress. This delay meant that the capital city of Saint-Pierre on the French island of Martinique was totally destroyed, and 36,000 people were killed.

Where possible, aid agencies try to reach the victims of a disaster as quickly as possible, with food and medical supplies.

When aid agencies first arrive on the scene they need to assess the situation as quickly as possible. Their first priority is to try and find survivors using specialized equipment and sniffer dogs. They will also need to provide the survivors with shelter, food, blankets, and clothes.

It is an enormous task and it is one that has to be done very quickly. It is estimated that 90 percent of earthquake victims die within the first 24 hours. Their task is made even harder if the disaster covers a large area or is in a mountainous region. This happened in Kashmir in 2005, when many survivors were stranded in remote areas for months without adequate food, shelter, and medical supplies.

The 2004 tsunami: can there even be too much aid?

On December 26, 2004, the world heard the shocking news that a tsunami had struck the coast of Thailand. The death toll was believed to be as high as 280,000. Due to the presence of many western tourists spending Christmas in the area, news of the disaster spread very quickly around the rest of the world. A colossal international rescue operation swung into action. This was supported by huge donations given both by the public and governments.

This caused a lot of controversy as so many other parts of the world are desperate for help. AIDS and malaria kill 280,000 Africans each month, but they receive no such assistance.

A huge relief effort was needed to help the stranded survivors of Hurricane Katrina in 2005.

Days later

In the days after a natural disaster it is essential to make sure that there is someone to take control. Shock often turns to desperation and anger, and desperate people will often fight over relief supplies if they are not given out in an orderly fashion. It might mean bringing in the army to stop any looting and fighting.

As the days pass, the chances of finding people alive lessen, but it is essential that the rescue teams keep looking. It has been known for people to be found alive in collapsed buildings two weeks after an earthquake.

In situations where there have been many deaths, there is also the possibility of disease, especially if the water supplies are polluted.

Dealing with the dead can present a major problem. Bodies need to be identified, and yet they need to be buried quickly and safely. After the tsunami in Thailand, thousands of bodies were preserved in refrigerated containers.

Relief supplies need to be distributed in an orderly fashion to keep a disaster situation under control.

Another after-effect of a major disaster is the destruction of food crops. This may mean a food shortage not just in the immediate future, but for several years after. In the past, famine following a natural disaster has taken more lives than the disaster itself.

Today, thanks to international agencies, food supplies can arrive quickly by plane or ship. These supplies will continue until the crisis is over. Sometimes this relief comes too late, as in the disaster in Niger 2005, when the situation became desperate before adequate help arrived.

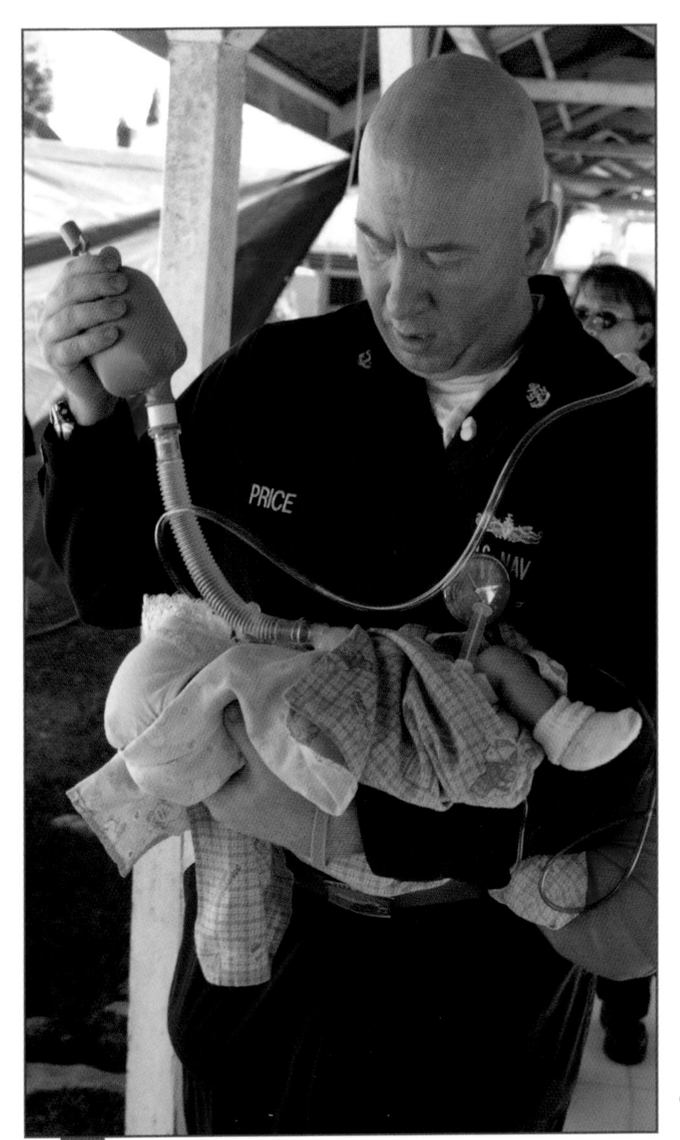

Military personnel can provide vital emergency aid.

Missing persons

In a major disaster, such as an earthquake in a city, some people simply disappear. No one knows whether these people are dead or alive. Families desperately search hospitals, camps, and mortuaries in a desperate bid to find their loved ones. Often babies will become separated from their parents and cannot be identified. This is particularly difficult if the disaster is spread over a large area. In many cases telephone cables will be destroyed, making communication really difficult.

Of course happy stories do emerge out of tragedy, when loved ones are reunited. Sometimes, however, the fate of the missing is never known.

Rich and poor

Disasters affect people in different ways. In richer countries where communication systems are more advanced, the media can make sure that everyone becomes aware of the crisis as soon as possible. Advanced transportation systems also mean that the injured can be tended to more quickly and supplies can be brought in. In contrast, poorer countries do not have the same facilities and require outside help. Sometimes this can go wrong, as in the case of Hurricane Katrina.

Hurricane Katrina: an avoidable disaster

Hurricane Katrina hit the southern coast of the U.S. with terrifying force in September 2005. The city of New Orleans was flooded as the flood defenses failed to cope with the amount of water. Hundreds died and nearly half a million people had to abandon their homes.

This disaster could have been avoided had the warnings been heeded. Now the U.S. faces a bill of at least $200 billion, as well as long-term hardship for many people who were already very poor.

REBUILDING LIVES

Humans are very adaptable. Despite many natural disasters, we have survived and rebuilt our lives. Hopefully we can all learn something from these disasters. One thing is certain though, anyone who has experienced one, will see the forces of nature in a different light.

Hurricane Mitch

Hurricane Mitch was one of the most destructive hurricanes on record. It tore through Central America in 1998, turning streets into lethal rivers. Towns, roads, and bridges were destroyed along with a lot of valuable farmland. Even worse, 18,000 people were killed. Central America is still reeling from the effects of this hurricane.

Recovery

When all the rescue workers have gone and the survivors are standing in the middle of all the devastation, they know they have an immense task ahead of them. They need to rebuild essential systems on which their community depends—schools, hospitals, essential services, roads, and communication systems to name but a few. None of this can happen though unless the people can recover their livelihoods. They need jobs, stores, healthcare, and schools for their children. What happens, though, if the disaster was too great for the people to recover?

Another obstacle is that there may simply not be enough people left. Many will have died and even more are likely to be refugees who never go back.

Hidden damage

Many victims will suffer mentally from the shock of a natural disaster. Many survivors literally lose everything that reminded them of who they were. Some people will be able to cope with this, while others will never recover—they have quite simply lost their identity.

Professional counselors and religious groups may be able to offer some form of comfort. To many victims though, the losses caused by the disaster have caused them to lose any religious beliefs they may have had.

Montserrat

Although it was thought to be dormant, the volcano called Souffrière Hills on the Island of Montserrat began to erupt in 1995. In 1997 it destroyed the capital, Plymouth. The British government, who were responsible for this small island, felt they had no option but to evacuate the population. Two-thirds of the 11,000 population left, most going to live in Britain. Many would like to return home, but the future still looks uncertain.

The ones to watch
Some of the world's volcanoes still cause concern:

• Vesuvius, Naples, Italy. Many believe an eruption is about to happen. Famous for its destruction of Pompeii in AD79.

• Popocatépetl, Mexico. Recently been showing increased activity.

• Cumbre Vieja, La Palma, Canary Islands (*below*). Threatens to cause a landslide, which could trigger a mega-tsunami.

Predicting a volcanic eruption

Monitoring stations constantly listen for any signs of movement (seismic activity) that might indicate the start of an eruption.

Satellites are also used to get information from sensors installed on volcanoes.

Monitoring equipment at sea can help issue warnings about a tsunami. If broadcast quickly enough, people living on the coast may have time to retreat to high ground. A tsunami warning system was installed in the Indian Ocean in 2006, to prevent another tragedy like the one that occurred in 2004. All too often, precautions are only taken after a major natural disaster has taken place.

Earthquakes on land continue to be the most unpredictable of all natural disasters.

Minimizing the impact

Even if you live in an earthquake zone, there are now programs to teach people how to improve their chances of survival. Modern buildings are designed to be able to withstand an earthquake, for example, by incorporating steel supports that will stop ceilings from collapsing.

Education plays an important part in minimizing the impact of a natural disaster. A 10-year-old British girl saw the signs that the tsunami was coming in 2004, and encouraged others to flee the coast. Many people foolishly believed it would be safer to stand on the beach, leaving them exposed to danger. The same thing happened in Lisbon in 1755.

If everyone who lives in disaster areas has the right knowledge and understanding, many lives may be saved.

The UN has a department called the International Strategy for Disaster Reduction (UNISDR). It collects and spreads information about how to reduce the impact of disasters, and installs warning devices.

Emergency planning

Local authorities take measures that are appropriate for their area.

Disaster-proof buildings

When San Francisco was hit by an earthquake in 1906, the houses (mainly wooden) stood up fairly well. However, fires caused by stoves and burst gas pipes destroyed the city. The lesson learned here was that buildings with some kind of flexibility can withstand a disaster. In poorer areas of the world, where houses are often badly built, many deaths occur.

The film called *Twister* (1996) was a story about people chasing storms. It also featured storm shelters which a lot of people in the U.S. use to protect themselves. Modern shelters are small concrete bunkers dug into the ground.

This NASA satellite image shows deep ocean tsunami waves about 20 miles (35 km) from Sri Lanka's southwestern coast.

For example, in areas prone to hurricanes or tornadoes, they can protect their citizens by building storm shelters and installing sirens.

Many Caribbean islands have hurricane shelters built of concrete and steel. Islanders can go in them when a storm is predicted. Some U.S. towns have similar shelters.

Getting people to safety requires special planning. In recent years, evacuation schemes have helped saved countless lives. During a major tropical cyclone in 1997, for instance, some 500,000 people were saved by emergency measures. Only 67 people died.

Is Naples ready for Vesuvius?

The city of Naples lies at the foot of Mount Vesuvius. The people who live there are well aware that the volcano will, one day, erupt again.

Vesuvius is the kind of volcano which will produce vast quantities of ash and rocky debris that will fall on the surrounding cities. It may also produce lethal clouds of gas.

At least 500,000 people live in the danger zone, but how do the authorities get them to evacuate? What if they refuse to leave?

Preventative measures

Local authorities also have the job of building defenses against floods and fires. In addition, they need to train their emergency services to cope with major disasters. As Hurricane Katrina proved, it is essential that this work is done at both local and international level.

Controlling disease

The same principles apply to disease—particularly diseases that spread following a disaster. The first priority here must be prevention. To minimize the risk of disease spreading, it is essential to provide clean water supplies and adequate sanitation. The bodies of those killed need to be disposed of quickly for the safety of others.

To reduce the risk of a pandemic (worldwide epidemic), medical research and spotting the first signs are very important. Vaccines can be provided for those people who are most at risk. Governments need to make sure that they have enough vaccines ready, or be able to produce them quickly.

Once again education plays a very important role. Simple hygiene, such as hand washing, can help prevent the spread of a disease. Also, it is important to make sure that people are drinking uncontaminated water.

When an epidemic does occur, nations have to be ready to fight it.

Health education is vital to minimize the risk of people catching deadly diseases.

This might mean calling in international experts, isolating the victims, and closing down travel routes (for example, airports) through which the infection might spread. Above all, governments need to be quite open about an outbreak, unlike China in the SARS outbreak (*see page 13*).

Polio makes a comeback

Polio is a dangerous, infectious disease which causes infant paralysis. It was hoped that with a rigid immunization program, this disease would be wiped out. Unfortunately, polio has recently made a comeback in Africa, Yemen, and Indonesia, where Islamic leaders have discouraged immunization programs.

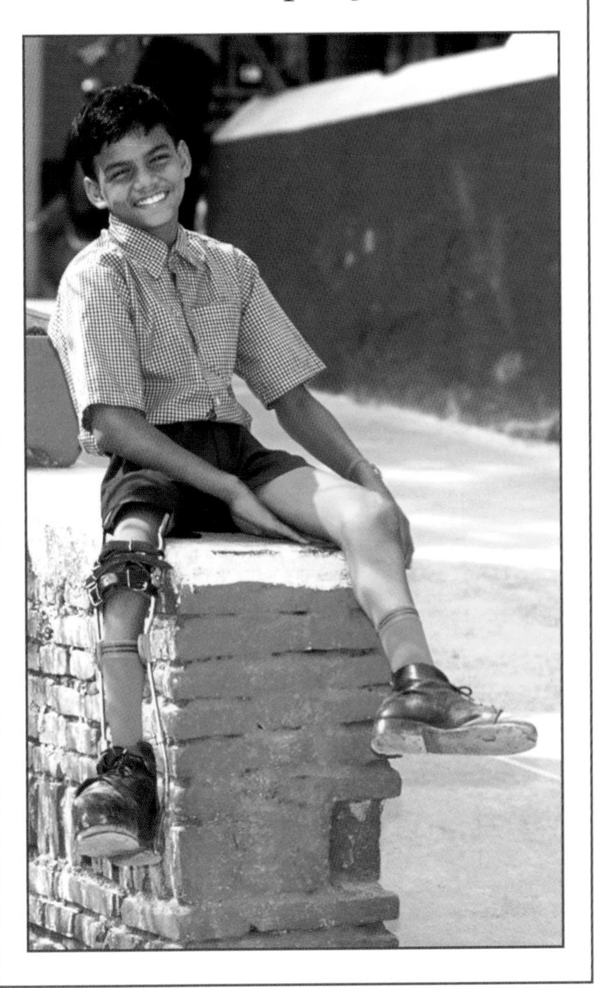

Could hospitals cope with a pandemic?

CALCULATING THE COST

It is totally impossible for any government to be totally prepared for a natural disaster. Because new strains of viruses break out all the time, there is no guarantee that the anti-viral drugs will work. This is particularly true of influenza. If a major flu epidemic broke out, hospitals would quickly be overwhelmed by the demand for beds.

It is essential to balance the risk of being always prepared for something that may not happen. Most nations that suffer from earthquakes, do not have the funds for a fleet of rescue helicopters to be on standby. As much as we can try and predict natural disasters, it would appear that nature always has the upper hand.

MAN'S EFFECT ON NATURE

Humans are the dominant species on Earth. We are ever more greedy for food and products and it is having a devastating effect on nature. By catching fish to eat and destroying forests, we are polluting the natural world. These actions are actually encouraging disasters to happen.

MISMANAGING NATURE

The deserts of the world are getting bigger all the time, and this is not helped by man's behavior. By using too much water, chopping down too many trees, and overgrazing already sparse pastures, we are risking the livelihoods of farmers that live on the edges of the desert. This mismanagement of land is nothing new. When prairie grasslands were plowed up to grow crops in the American Midwest, the soil turned to dust from lack of rain.

Sand storms often happen where deserts are encroaching on farmland.

The land became infertile and thousands of people died from starvation and lung diseases caused by the dust. A similar thing is happening in the Amazon rainforest in Brazil. People are now trying to regenerate farmland in desert areas and to replant forests.

When forests on hills and mountains are cut down, this can cause flooding. The soil is unable to cope with the amount of water. Bangladesh, for example, is affected by repeated flooding caused by rainwater formed in the Himalayas. Dams and levees have been built to reduce the risk of flooding in affected areas, but it is not always wise to interfere with the natural flow of a river.

The Aral Sea disaster

The Aral Sea that borders Kazakhstan and Uzbekistan was once a large and healthy sea. However, in the 1960s, the Soviet government diverted water from the rivers that fed the Aral Sea to irrigate their cotton fields. As a result, the sea shrunk. The water also became so salty it killed the fish. This is one of the worst ecological disasters caused by man.

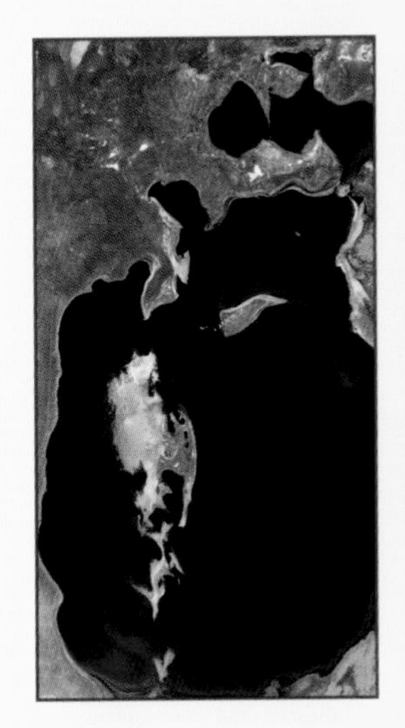

1989

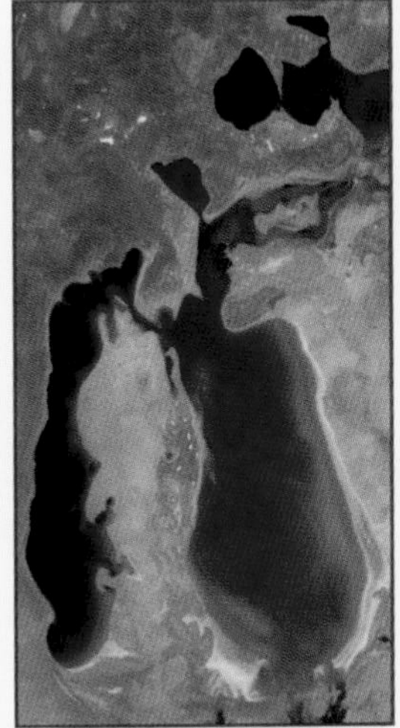

2003

In 1993, the Mississippi and Missouri Rivers flooded after months of heavy rain.

Nature as a garbage can

Man continues to pollute nature with waste products. Waste pollutes the rivers, destroys river life, and makes the water unsafe to drink.

Chemical fertilizers used on crops can also end up in our rivers and seas. This creates a toxic algae which poisons fish and destroys coral reefs.

Virtually all life in the water off the coast of Shanghai in China has been destroyed by pollution from the Yangtze River. The same thing is happening in the summer off the coast of Louisiana.

Destructive gases

Gases released into the air, such as chlorine and bromine, are damaging the ozone layer. The ozone layer helps to protect our planet from the harmful rays of the sun that cause cancer.

Industrial waste pollutes natural water sources.

Is global warming really happening?

"Global warming" describes an increase in the average temperature on Earth that can cause climate change.

One major effect of these rising temperatures is the melting of polar ice caps. The Arctic ice cap has already shrunk significantly and there is a real threat of future worldwide flooding.

Scientists now think that our actions contribute to global warming. We need to take action now and look after our planet. If we wait for proof that what we are doing is harmful, we may find that it is too late to do anything about it.

Large holes have already been detected over the Antarctic.

Fossil fuels—coal, oil, and gas—are causing the most concern. We burn huge amounts of these every day to run our power plants, factories, airplanes, and even our cars. By burning these fuels we release vast quantities of carbon dioxide (CO_2) into the air. Scientists believe this is causing "global warming." Global warming is being linked to some natural disasters. For example, as sea levels rise with the melting of the polar ice caps, many nations are at risk of flooding. Sea levels have already risen by almost six inches (15 cm) in the last 10 years.

There is a threat of flooding from melting ice caps

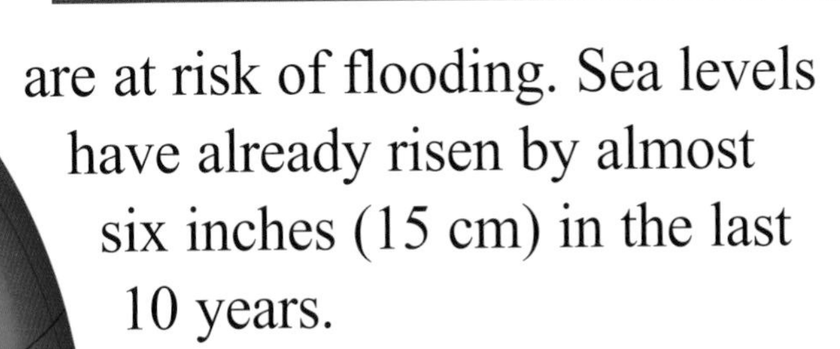

The Thames Barrier in London

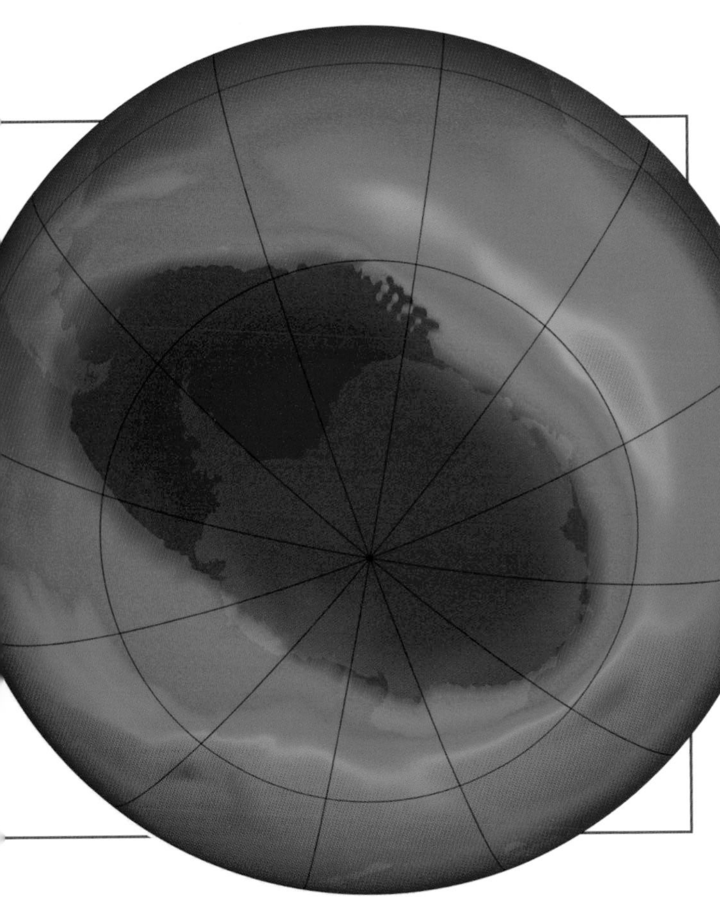

As global temperatures rise, it is thought that tropical cyclones will become more frequent and more powerful. This proved true in the case of Hurricane Katrina in 2005, quickly followed by Hurricanes Rita and Stan.

Encouraging disease

Human behavior can also trigger disease and epidemics. The danger of bird flu has increased because we keep vast numbers of chickens in battery farms. Today, as people travel farther and more often, the risk of a new outbreak of bird flu causes major concern. Any new strain could quickly be carried

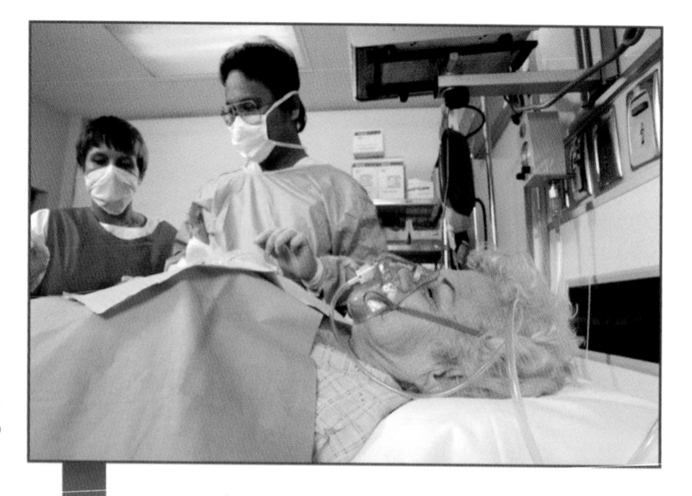

Modern medicine has made great advances

around the world by tourists and business travelers.

Messing around with nature can have serious consequences. A disease which affects cattle called BSE or "Mad Cow Disease" was probably caused by farmers feeding animal-based products to their cattle. Cattle are vegetarian and are not used to eating this type of food. BSE is believed to be the cause of a fatal and incurable brain disease in humans, called CJD. The fear of a widespread epidemic meant that more than a million cattle had to be slaughtered.

NEW THREATS

Man may be taking risks with nature in other ways. Genetic modification—altering the gene structure—to try and improve crop yields may have some very unpredictable side effects.

The world also experiments with biological weapons. These are designed to spread diseases such as anthrax, Ebola, smallpox, or the bubonic plague. Any use of these, or even an accident, could have disastrous consequences.

During the Vietnam War in the 1960s, the U.S. military forces used chemicals to strip the leaves off trees. This was to expose their enemy, the Vietcong, who used the trees as cover.

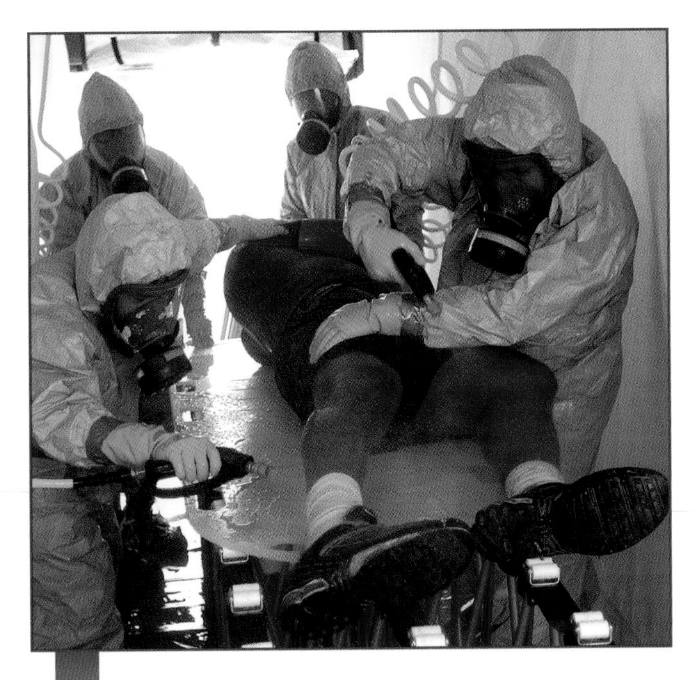

An emergency drill to prepare for biochemical attacks

Since then, at least 150,000 children in Vietnam have been born with severe deformities. Many adults—both Americans and Vietnamese—have suffered serious health problems associated with the use of these chemicals.

Finding the balance

The international community is now trying to make everyone aware of the damage they are doing to the planet. A program called the Kyoto Protocol, set up in 1997, was designed to help us all reduce our emissions of CO_2.

However, not everyone is convinced that it is necessary to take such measures. Many countries are still reluctant to sign up to the Kyoto Protocol.

Finding the right balance between man and nature is going to take a lot more work, if we are going to get people to change their way of life.

As our population continues to rise at such an alarming rate, natural disasters are only one small part of the danger posed by man.

Nations gather in Japan in 1997 to discuss environmental problems affecting our planet

THE FUTURE

Humans will never be free from the threat of natural disasters. We can try and learn from what has happened in the past, but we never know what the future holds. In the distant past there have been catastrophic events far, far greater than anything humans have ever witnessed. Could a disaster of this magnitude strike again?

MEGA-DISASTERS

There have been volcanic eruptions in the past that are far larger than any disaster in recent history. About 74,000 years ago, Mount Toba in Sumatra, Indonesia, erupted with an enormous force. It was probably the biggest eruption in two million years. Scientists have estimated that as many as 75 percent of all plant species in the Northern Hemisphere died as a result. Man may have been pushed to the edge of extinction.

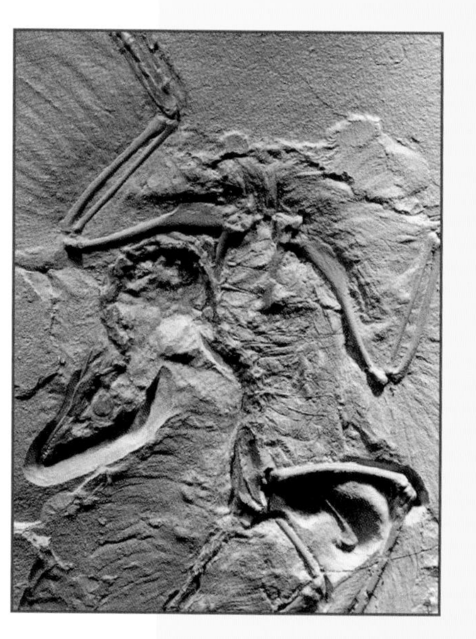

The great extinctions of the past

The greatest case of extinctions due to climate change came at the end of the Permian period, 250 million years ago. Almost 90 percent of all life forms were made extinct. At the end of the Cretaceous period, 65.5 million years ago, about half of all life-forms, including dinosaurs, were lost forever.

The Yellowstone super-volcano

Yellowstone National Park, is famous for its bubbling geysers and hot and brilliant-colored mineral pools. Although these are very pretty, they are just surface wonders of a huge volcanic force that lies beneath. This force is building up, and gradually pushing the land into an enormous dome. Could Yellowstone turn into a "super-volcano?"

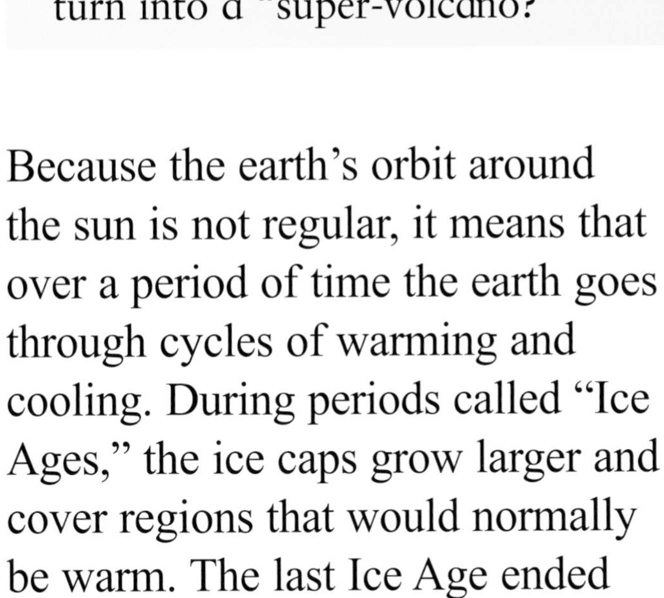

Because the earth's orbit around the sun is not regular, it means that over a period of time the earth goes through cycles of warming and cooling. During periods called "Ice Ages," the ice caps grow larger and cover regions that would normally be warm. The last Ice Age ended 10,000 years ago.

According to the known cycles, Earth is due for another Ice Age. However, the fact that Earth appears to be getting gradually warmer, points to the fact that it is man, not nature, that is responsible for the recent increases in temperature on planet Earth.

Apocalypse now

Some scientists are predicting that if we continue to fill the air with carbon dioxide, the reduction in essential oxygen will cause a massive extinction.

"Superbugs"—dangerous bacteria or viruses resistant to treatment—could also emerge and cause a worldwide pandemic. If this were to happen the entire human race would be endangered.

In the past, Earth has been hit by meteorites and asteroids hurtling from outer space. Are we in danger of another impact?

It is believed that large asteroids have hit Earth in the past, destroying the habitat and causing the extinction of many animals. A crater in Mexico suggests that Earth was hit by an asteroid that measured 6 miles (10 km) in diameter, around 65 million years ago. An impact like this would cause an explosion many millions of times more powerful than any atomic bomb. It would alter Earth's atmosphere, plunging the planet into a state of prolonged winter. It is now believed that dinosaurs became extinct due to an asteroid collision.

This crater in Mexico might be from an asteroid that hit Earth about 65 million years ago

Future collisions

There are several hundred "near-Earth" asteroids in space, which could hit our planet with little warning. One called 4581 Asclepius, measuring 980 ft (300 m) in diameter, missed Earth by 250,000 miles (400,000 km) in 1989. Astronomers estimate that collisions with smaller asteroids take place approximately every ten million years. The chances of this happening in our lifetime are therefore remote.

Humans have always loved to speculate about the destruction of planet Earth. Shortly after World War II, an all-out nuclear war seemed to threaten the human race.

Will we never learn?

Why do people continue to live in areas where natural disasters frequently occur? The cities of San Francisco, Los Angeles, and Tokyo have already been devastated by disasters, and yet people continue to take the risk.

One reason is that people have lived there for a long time. The areas are set up to sustain human life and people feel it is easier to stay and repair the damage than to move to an unfamiliar place.

In some parts of the world, measures are being taken to reduce the threat of natural disasters. Buildings to resist earthquakes, early warning equipment, and sea defenses like the Thames Barrier in London are just a few examples.

However hard we try to fight nature, natural disasters will always happen, and the likelihood is they will probably get worse.

Today's vision of man's destruction is global warming.

LIVING WITH NATURE

For the time being we can only deal with what we know on a day-to-day basis. We have to accept that natural disasters will continue to happen; there is no way of avoiding them. We need to learn how to minimize their impact. We have recovered from disasters in the past, and we will continue to do so in the future. However, we must learn from what has happened in the past and try to avoid unnecessary disasters. Only by watching and learning from our natural world, will we ever come close to living in harmony with it.

CHRONOLOGY

c.72,000 BC—Mount Toba in Indonesia erupted.

AD 79—Mount Vesuvius, Italy erupted and destroyed Pompeii.

1556—Earthquake in China, killed almost 830,000 people.

1755—An earthquake followed by a tsunami destroyed Lisbon.

1780—"The Great Hurricane" struck Martinique, Barbados.

1815—Mount Tambora, Indonesia, erupted killing 12,000.

1826–37—The Second Cholera Pandemic spread around the world from India, killing millions.

1857—107,000 people were killed by an earthquake in Japan.

1887—The Huang He (Yellow River) in China flooded, killing 1.5 million people.

1896—A tsunami hit the coast of north-east Honshu, Japan, killing 27,000.

1900—12,000 people killed in a hurricane at Galveston, Texas.

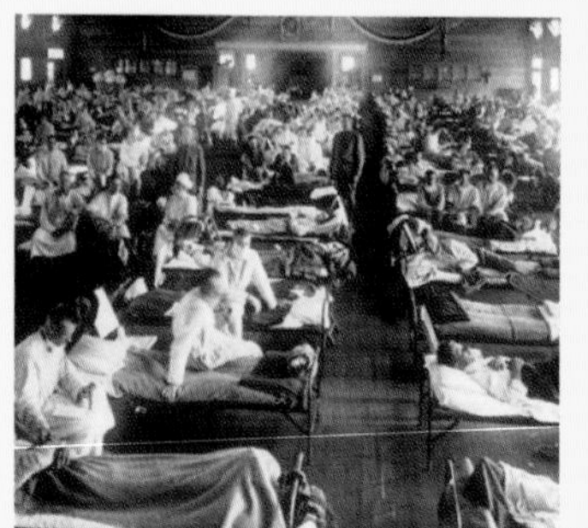

1902—Mont Pelée in Martinique erupted, killing most of the population.

1911—A flood on the Yangtze River killed 200,000 in the region of Shanghai, China.

1918–19—The "Spanish flu" pandemic killed at least 25 million people worldwide.

1920—180,000 people died in an earthquake in Xining, Gansu Province, China.

1927—"Great Mississippi Flood."

1931—Huang He (Yellow River) and Yangtze River flooded, killing three million people in China.

1932–40—Prolonged drought resulted in the "Dust Bowl" in Midwest America.

1960—Great Chilean Earthquake and resulting tsunami that hit Hawaii killed 3,000 people.

1963 and 1970—Tropical cyclones hit Bangladesh.

1976—Worst earthquake of modern times hit Tangshan, Hebei Province, China.

1984–5—Drought in East Africa.

1988-89—Bangladesh suffered its worst flood on record.

1993—Mississippi and Missouri Rivers flooded.

1998---Hurricane Mitch devastated Honduras and Nicaragua, killing 18,000.

2003—An earthquake killed 26,000 in Bam, Iran.

2004—An earthquake followed by a tsunami in Indonesia, Thailand, Sri Lanka, and India.

2005—Hurricane Katrina devastated New Orleans and surrounding areas.

2006—Danube River flooded.

2008—A devastating cyclone in Myanmar (Burma) and earthquake in China caused huge losses of life.

ORGANIZATIONS AND GLOSSARY

CARE—Cooperative for Assistance and Relief Everywhere. Founded in the U.S. in 1945, but is now based in Brussels.

European Union (EU)—Now a group of 27 European countries, it was formed to promote peace through trading and political agreements.

Food and Agriculture Organization (FAO)—Specializes in food production, particularly in the Third World.

Médicins Sans Frontières (MSF)—Also known as Doctors Without Borders. Volunteer doctors and medics, supported by donations, travel the world.

Office for the Coordination of Humanitarian Affairs (OCHA)—Deals with natural disasters and other emergencies.

Oxfam International—This major charity was founded in 1942 as a committee responsible for famine relief.

Red Cross—Plays a leading role in relief work after natural disasters, and gives assistance to the sick and wounded.

Save the Children—Founded in Britain in 1919, works in more than 110 countries focusing on the needs of children.

United Nations (UN)—An international organization founded in 1945 to promote peace, security, and economic development.

World Food Program (WFP)—Distributes emergency food to the victims of natural disasters.

World Health Organization (WHO)—Deals with infectious diseases and epidemics and global health issues.

Asteroid—A lump of rock or ice in space, more than 160 ft (50 m) in diameter.

Avian flu—A deadly type of flu virus that infects birds.

Cyclone—A circulating pattern of air around an area of low pressure.

Delta—Flat area at the mouth of a river where the main stream splits up into several branches.

Earthquake—The sudden violent shaking of the ground.

Epicenter—The point on the earth's surface directly above the focus of an earthquake.

Fossil fuels—Most of our main fuels for making energy—oil, natural gas, and coal.

Meteoroid—Lump of rock or ice in space, less than 160 ft (50 m) in diameter. If it enters Earth's atmosphere it is called a meteor. If it reaches the ground it is called a meteorite.

Monsoon—A wind that blows across southern Asia at different seasons. It brings heavy rains which often cause flooding.

Richter scale—The scale by which earthquakes are measured; named after its inventor, the American seismologist Charles Richter.

Seismic—Describes any shaking of the earth's surface.

Seismology—The word for the scientific study of earthquakes comes from the ancient Greek for an earthquake, *seismos.*

Tectonic plate—The massive, moving plates into which the earth's crust is divided are called tectonic plates.

Tsunami—A massive wave caused by an undersea earthquake or volcanic eruption.

INDEX

Photo Credits:
Abbreviations: l-left, r-right, b-bottom, t-top, c-center, m-middle. Front cover main – Comstock. Front cover ml, 10bl, 16tl, 34tr, 36mr – Digital Vision. Front cover c, 1m, 1r, 2ml, 2bl, 4tl, 8-9, 9br, 23tr, 28mc, 37tr, 38ml, 40tl, 42tl – Corbis. Front cover mr, 14ml – Win Henderson/FEMA. 1l, 3tr, 12tr, 15tr, 15br, 17tr, 20mb, 21tr, 25tr, 27bl, 39tl, 45br – U.S. Navy. 2-3, 6m, 7bl, 41tr – Photodisc. 4-5, 10mr – World Food Programme. 5tr, 15tl, 17bl, 22mr, 26tr – Marty Bahamonde/FEMA. 6tr, 32mr, 33bl – WHO/P. Virot. 8mr – Library of Congress. 11br – Mark Wolfe/FEMA. 12br – Myles Chilton/WPN. 14tr, 35bl, 43mr, 45tl – Andrea Booher/FEMA. 14bl – Barry Voight. 15bl – ACT. 16br – REUTERS/Pierre Holtz. 18bl, 27br, 34br – Corel. 19mr – David Rydevik. 20tl – Jocelyn Augustino/FEMA. 22br – Marvin Nauman/FEMA. 24br – Mark Lawhead. 28tr – www.istockphoto.com/Matt Matthews. 29tr – Disasterman Ltd. 30br – www.istockphoto.com /Lou Oates. 31tr, 35ml, 35mr, 37bl – NASA's Earth Observatory. 31br – www.istockphoto.com/Kate Shephard. 33tr – Michael Rieger/FEMA. 37mr – www.istockphoto.com/Chris Schmidt. 39br – © European Community. 40mb – www.istockphoto.com/Bob Ainsworth. 42m – www.istockphoto.com. 44tr – www.istockphoto.com/Bill Gruber. 44m – Courtesy of the National Museum of Health and Medicine, Armed Forces Institute of Pathology, Washington D.C.